ZOO

Farzana Sarup
Edited by Bonnie Dobkin

Wonder House

AT THE ZOO

I love to see the animals
When I go to the zoo.
I visit all of them and learn
What each one likes to do.

I see big cats and crocodiles,
Giraffes as tall as trees.
I smile at them and sometimes
I think they wink back at me!

TIGER

I am a handsome tiger,
I have orange stripes and black.
Right now I'm feeling hungry so
You'd better stand way back!

Because I am **NOCTURNAL**
I hunt for food at night.
With my big teeth and long sharp claws
I am a scary sight!

LION

Hello there! I’m a lion.
Don’t you love my handsome mane?
I think that I’m good looking.
(But please don’t think that I’m vain.)

We lions live together
In large groups that we call **PRIDES**.
We take care of each other
With our cubs close by our sides.

And all of us are **CARNIVORES**
That means we just eat meat.
So we go hunting as a group
To catch each yummy treat.

What are young lions called?

HIPPOPOTAMUS

I am a hippopotamus
You see that I'm well fed.
I have short legs, a tiny tail,
A wide mouth and big head.

You may think that I'm hiding
Since I'm often out of sight.
But I'm just **SUBMERGED** in water
Keeping cool until it's night.

When evening comes I leave my pond
And climb out on the land.
I munch on grass for hours and hours
'Til I can hardly stand!

ELEPHANT

I am a giant animal,
The biggest one on land!
I have large ears and two white **TUSKS**.
I really am quite grand.

My trunk is very special—
It helps me drink and smell.
But I can use it to move logs
Or hug my calf as well!

How else could an elephant use its trunk?

Hello! I am a crocodile.
Know what I do for fun?
Like all **REPTILES** my blood is cold.
So I lie in the sun!

I have rough skin and sharp, sharp teeth
My bite is fierce and strong.
I use my tail to help me swim—
Say, want to come along?

ZEBRA

We zebras look like horses.
Most people know us well.
Our stripes are black—or are they white?
It's really hard to tell.

We love to **GRAZE** on grass all day.
We eat it to the ground!
That's why my zebra herd and I
Must always move around.

Why do you think a zebra has stripes?

What kind of food do you think giraffes eat?

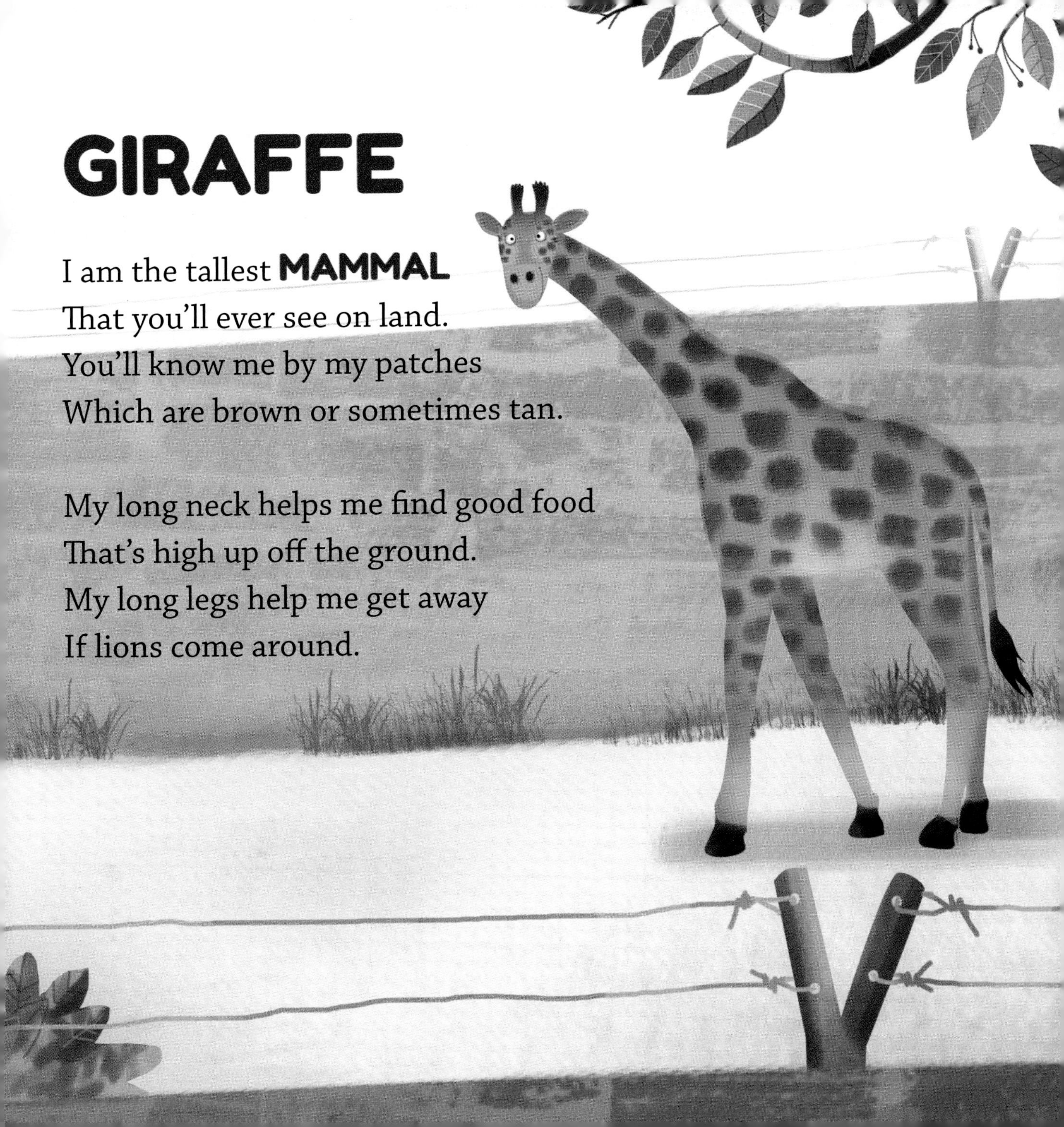

GIRAFFE

I am the tallest **MAMMAL**
That you'll ever see on land.
You'll know me by my patches
Which are brown or sometimes tan.

My long neck helps me find good food
That's high up off the ground.
My long legs help me get away
If lions come around.

DOLPHIN

I am a clever mammal.
I'm as playful as can be!
The next time you are at the zoo
Please come and visit me!

I'll show you how fast I can swim
And how high I can leap.
I also can **COMMUNICATE**
With whistles, clicks, and squeaks.

I breathe air through the blowhole
You can see upon my head
I love to eat fish, shrimp, and squid
Whenever I am fed.

How do dolphins talk?

PENGUIN

Some think that I'm a funny bird
I waddle when I walk.
I live in water half the time
On land I scoot and squawk.

Since I have flippers and not wings
You'll never see me fly.
My feathers, which are **WATERPROOF**
Help keep me warm and dry.

Why don't
penguins
get cold?

CHOOSE THE BEST WORD

TIGER

☐ Tusk ☐ Mane ☐ Stripes ☐ Short legs

PENGUIN

☐ Flippers ☐ Mammal ☐ Tail ☐ Large ears

CROCODILE

☐ Blowhole ☐ Fur ☐ Calf ☐ Rough skin

ELEPHANT

☐ Cubs ☐ Tusks ☐ Herd ☐ Nocturnal

LION

☐ Pride ☐ Giant ☐ Trunk ☐ Whistle

GIRAFFE

☐ Reptile ☐ Pond ☐ Scary ☐ Long neck

LET'S TALK!

- When does a tiger like to hunt?
- What are young lions called?
- Why are a hippo's eyes and nose near the top of its head?
- How else could an elephant use its trunk?
- Why do you think a crocodile has sharp teeth?
- Why do you think a zebra has stripes?
- What kind of food do you think giraffes eat?
- How do dolphins talk?
- Why don't penguins get cold?